WIRELESS TECHNOLOGY
from Then to Now

BY RACHEL GRACK

AMICUS | AMICUS INK

Sequence is published by Amicus and Amicus Ink
P.O. Box 1329, Mankato, MN 56002
www.amicuspublishing.us

Library of Congress Cataloging-in-Publication Data
Title: Wireless technology from then to now / by Rachel Grack.
Description: Mankato, Minnesota : Amicus/Amicus Ink, [2020] | Series: Sequence developments
 in technology | Includes bibliographical references and index. | Audience: Grade 4 to 6.
Identifiers: LCCN 2018048680 (print) | LCCN 2018050791 (ebook) | ISBN 9781681517667
 (pdf) | ISBN 9781681516844 (library binding) | ISBN 9781681524702 (pbk.)
Subjects: LCSH: Wireless communication systems--History--Juvenile literature.
Classification: LCC TK5102.4 (ebook) | LCC TK5102.4 .K64 2020 (print) | DDC 621.38209--dc23
LC record available at https://lccn.loc.gov/2018048680

Editor: Wendy Dieker
Designer: Aubrey Harper
Photo Researcher: Holly Young

Photo Credits: KOHb/iStock cover; Masarik/Shutterstock cover; OJO Images Ltd/Alamy 4;
Popular Science Monthly, Volume 45, 1894, Author Unknown/WikiCommons 7; The Complete
Radio Book, The Century Co., New York, p. 32, Raymond Francis Yates, Louis Gerard Pacent
(1922), Author Unknown/WikiCommons 7; Universal Images Group/Getty 8; Bettmann/Getty
11, 19; Archive Photos/Getty 12; Daderot/WikiCommons 12–13; Windell Oskay/WikiCommons
14–15; Classic Stock/H. Armstrong Roberts/MaryEvans 16; Corbis Historical/Getty 20–21; Jose
Luis Pelaez Inc/Getty 23; Suzy Bennett/Alamy 24; MaestroBooks/iStock 27; jamesteohart/iStock
28–29

Printed in the United States of America

HC 10 9 8 7 6 5 4 3 2 1
PB 10 9 8 7 6 5 4 3 2 1

TABLE OF CONTENTS

Radio Waves all Around	5
On Air	9
More Power	13
Mobile Wireless	18
Cutting Wires	22
Into the Future!	26

■ ■ ■ ■ ■

Glossary	30
Read More	31
Websites	31
Index	32

Cell phones are devices that use the radio waves flying in the air around us.

Radio Waves all Around

You call your friend on a cell phone. How does it carry your voice? It uses radio waves. These invisible waves are all around us. They help people and devices communicate by sending signals through the air. Wireless technology started with a spark in 1886. Since then, we keep finding ways to use these **airwaves**.

In 1886, scientist Heinrich Hertz crossed electricity with magnetic fields. Spark-covered waves zigzagged in his lab. He discovered radio waves! Hertz thought these waves were only useful for science experiments. He was wrong! Soon, other inventors found ways to send **data** on these airwaves.

Hertz watches electricity spark across invisible radio waves.

Heinrich Hertz discovers radio waves.

1886

...LOADING...LOADING...

Marconi first experimented with sending radio signals across his yard.

Heinrich Hertz discovers radio waves.

1886   1894

ADING . . . LOADING .

Guglielmo Marconi sends Morse code on radio waves.

On Air

In 1894, Italian inventor Guglielmo Marconi tried to get radio waves to carry sounds. He built a radio **transmitter**. It sent beeps of **Morse code** to a **receiver**. It worked! He worked to send messages farther and farther. In 1901, he sent sounds from England to Canada. That's 2,100 wireless miles (3,380 km)!

Marconi worked on distance. But other people wondered if radio waves could carry voices. In 1900, Canadian Reginald Fessenden spoke a message over the airwaves. "Is it snowing where you are?" he asked. His friend answered yes. These were the first words on air.

Fessenden spent his adult life inventing radio technologies.

Heinrich Hertz discovers radio waves.

Reginald Fessenden speaks the first words on air.

1886   1894   1900   G . . . L O A D I N G .

Guglielmo Marconi sends Morse code on radio waves.

De Forest's amplifier (inset) made it possible for families to gather around a radio to listen to shows.

Heinrich Hertz discovers radio waves.

Reginald Fessenden speaks the first words on air.

1886 1894 1900 1907

Guglielmo Marconi sends Morse code on radio waves.

Lee de Forest invents the vacuum tube.

LOADING...

More Power

In 1907, Lee de Forest found a way to make radio waves stronger. His vacuum tube was the first **amplifier**. People called it an amp for short. These glass tubes were used to build radios for everyone. Amps helped carry live radio shows into homes across North America. By 1930, families gathered around their radios. They listened to their favorite shows.

Vacuum tubes used a lot of power. They got very hot. In 1947, scientists at Bell Labs worked to find an amp that was small, sturdy, and stayed cool. They invented the **transistor**. This tiny invention was a big deal! It led to portable radios. People could tune in to wireless waves anytime and anywhere.

Bell Labs keeps the very first working transistor on display.

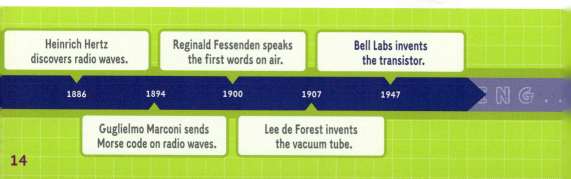

Heinrich Hertz discovers radio waves.	Reginald Fessenden speaks the first words on air.		Bell Labs invents the transistor.
1886	1894	1900	1907	1947
	Guglielmo Marconi sends Morse code on radio waves.	Lee de Forest invents the vacuum tube.	

A family gathers to watch TV. The picture and sounds came through the airwaves.

Heinrich Hertz discovers radio waves.

Reginald Fessenden speaks the first words on air.

Bell Labs invents the transistor.

1886   1894   1900   1907   1947   1950s

Guglielmo Marconi sends Morse code on radio waves.

Lee de Forest invents the vacuum tube.

Radio waves carry TV broadcasts into homes.

Meanwhile, inventors worked to send motion pictures through radio waves. Television! By the 1950s, many North American homes had **antennas** and TVs. Cameras turned pictures and sounds into signals. The signals rode on the airwaves to antennas on the homes. Then, the TV turned the signals back into pictures and sounds.

Mobile Wireless

Inventors kept finding ways for people to talk to each other over radio. Some cars had big radio phone systems installed. But not many phones could use the available radio waves at one time. In the 1970s, a system called Improved Mobile Telephone Service (IMTS) took calls on the road. This service allowed more users at a time. The equipment in the car was more compact, too.

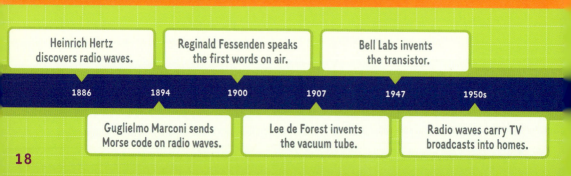

Heinrich Hertz discovers radio waves.	Reginald Fessenden speaks the first words on air.	Bell Labs invents the transistor.
1886	1900	1947
1894	1907	1950s
Guglielmo Marconi sends Morse code on radio waves.	Lee de Forest invents the vacuum tube.	Radio waves carry TV broadcasts into homes.

The first car phones were installed in the 1940s, but the service was limited.

LOADING...LOADING...

IMTS lets people make wireless calls from their cars.

Heinrich Hertz
discovers radio waves.

Reginald Fessenden speaks
the first words on air.

Bell Labs invents
the transistor.

1886   1894   1900   1907   1947   1950s

Guglielmo Marconi sends
Morse code on radio waves.

Lee de Forest invents
the vacuum tube.

Radio waves carry TV
broadcasts into homes.

At the same time, phone companies were working on a handheld mobile phone. In 1983, Motorola sold the first wireless handheld cell phone. They called it the DynaTAC 8000x. This early cell phone was the size of a sub sandwich. It wouldn't have fit in your pocket. But it was paving the way to the cell phones we have today.

A man uses one of the earliest handheld mobile phones.

Motorola offers the world's first handheld cell phone.

1970s 1983

ING...LOADING...

IMTS lets people make wireless calls from their cars.

Cutting Wires

People were sending data over long distances. But engineers were thinking about how to connect nearby electronics without wires. Created in 1994, **Bluetooth** technology uses radio waves. It can be used to connect an iPod to ear buds, for example. Bluetooth connects a controller to a gaming console. It can connect a smartphone to a car's audio system.

Bluetooth technology connects this game controller to the console without cords.

Heinrich Hertz discovers radio waves.		Reginald Fessenden speaks the first words on air.		Bell Labs invents the transistor.	
1886	1894	1900	1907	1947	1950s
	Guglielmo Marconi sends Morse code on radio waves.		Lee de Forest invents the vacuum tube.		Radio waves carry TV broadcasts into homes.

Motorola offers the world's
first handheld cell phone.

1970s   1983   1994   ...LOADING...

IMTS lets people make wireless
calls from their cars.

Bluetooth connects
devices wirelessly.

Heinrich Hertz discovers radio waves.		Reginald Fessenden speaks the first words on air.		Bell Labs invents the transistor.	
1886	**1894**	**1900**	**1907**	**1947**	**1950s**
	Guglielmo Marconi sends Morse code on radio waves.		Lee de Forest invents the vacuum tube.		Radio waves carry TV broadcasts into homes.

24

By the late 1990s, people were going online to the World Wide Web. Internet data came through cables and phone lines. But engineers knew there had to be a way to get online wirelessly. In 1997, Wi-Fi was invented. Soon computers were using radio waves to connect to the Internet.

People in an airport sit in a Wi-Fi area so they can connect computers to the Internet.

Motorola offers the world's first handheld cell phone.

Wi-Fi makes wireless Internet connections possible.

1970s   1983   1994   1997   ADING...

IMTS lets people make wireless calls from their cars.

Bluetooth connects devices wirelessly.

Into the Future!

More and more people started using wireless devices. The wireless **spectrum** became very crowded. Imagine thousands of devices trying to use the same waves at the same time. In 2011, scientists found they could spin radio waves. These twisted waves hold more signals. They hope to free up future wireless space.

Many people depend on wireless connections. Radio waves must carry big loads of data.

Heinrich Hertz discovers radio waves.	Reginald Fessenden speaks the first words on air.	Bell Labs invents the transistor.

1886   1894   1900   1907   1947   1950s

Guglielmo Marconi sends Morse code on radio waves.	Lee de Forest invents the vacuum tube.	Radio waves carry TV broadcasts into homes.

	Motorola offers the world's first handheld cell phone.	Wi-Fi makes wireless Internet connections possible.	
1970s	1983	1994	1997	2011
IMTS lets people make wireless calls from their cars.		Bluetooth connects devices wirelessly.	Scientists twist radio waves to make room for more data.

27

Heinrich Hertz discovers radio waves.

Reginald Fessenden speaks the first words on air.

Bell Labs invents the transistor.

1886   1894   1900   1907   1947   1950s

Guglielmo Marconi sends Morse code on radio waves.

Lee de Forest invents the vacuum tube.

Radio waves carry TV broadcasts into homes.

Fewer and fewer devices need wires to connect. Engineers keep looking for ways to make the wireless networks carry more data. In 2018, 5G wireless went live. These networks are the speediest connections. Could we make it a day without these invisible waves? Probably not. If Hertz only knew how useful radio waves would be!

If we could see radio waves, we'd see all the data moving around the city.

Motorola offers the world's first handheld cell phone.

Wi-Fi makes wireless Internet connections possible.

5G networks make wireless connections faster than ever.

1970s   1983   1994   1997   2011   2018

IMTS lets people make wireless calls from their cars.

Bluetooth connects devices wirelessly.

Scientists twist radio waves to make room for more data.

Glossary

airwave Another name for a radio wave.

amplifier A part inside an electronic device used to make radio waves stronger.

antenna A rod or wire used to send and receive radio waves.

Bluetooth A way to connect electronic devices with radio waves instead of wires or cables.

data Words or images stored and shared by electronics.

Improved Mobile Telephone Service (IMTS)
A mobile telephone service that connected landline phones with car phones.

Morse code An alphabet code in which letters are sent using groups of long and short beats of light or sound.

receiver An electronic device used to detect and take in radio waves.

spectrum The whole range of radio waves, from small waves to large waves.

transistor A type of amplifier in an electronic device used to make radio waves stronger.

transmitter An electronic device used to send radio waves.

Read More

Amstutz, Lisa. *Smartphones.* Mendota Heights, Minn.: North Star Editions, 2017.

Dinmont, Kerry. *Communication Past and Present.* Minneapolis, Minn.: Lerner Publications, 2019.

Slingerland, Janet. *Wi-Fi.* Lake Elmo, Minn.: Focus Readers, 2018.

Websites

Kids Geography | Electromagnetic Spectrum
https://kidsgeo.com/geography-for-kids/electromagnetic-spectrum

NASA Science | Radio Waves
https://science.nasa.gov/ems/05_radiowaves

Radio Wave Facts for Kids
https://kids.kiddle.co/Radio_wave

Index

amplifiers 13

antennas 17

Bell Labs 14

Bluetooth 22

car phones 18

cell phones 5, 21, 22

De Forest, Lee 13

discovery 6

Fessenden, Reginald 10

5G 29

Hertz, Heinrich 6, 29

Improved Mobile Telephone Service (IMTS) 18

Marconi, Guglielmo 9, 10

mobile phones 18, 21

Morse code 9

portable radios 14

radio shows 13

receivers 9

television 17

transistors 14

transmitters 9

twisted radio waves 26

vacuum tubes 13, 14

Wi-Fi 25

wireless devices 23, 26, 29

About the Author

Rachel Grack has worked in children's nonfiction publishing since 1999. Rachel lives on a small desert ranch in Arizona. She enjoys spending time with her family and barnyard of animals. Thanks to our wireless world, her ranch stays tapped into developing technology.